Y0-CVN-527

5
1/16
2/21

DATE DUE

THE CHICAGO PUBLIC LIBRARY

WEST PULLMAN BRANCH
800 W. 119TH STREET
CHICAGO, IL 60643

OCEANS

BIOMES OF NORTH AMERICA

Lynn M. Stone

Rourke
Publishing LLC
Vero Beach, Florida 32964

© 2004 Rourke Publishing LLC

All rights reserved. No part of this book may be reproduced or utilized in any form or by any means, electronic or mechanical including photocopying, recording, or by any information storage and retrieval system without permission in writing from the publisher.

www.rourkepublishing.com

PHOTO CREDITS: Pages 7, 8, 10, 21, 22 , cover © Lynn M. Stone; title page, p. 12, 15, 17, 18 © Brandon Cole

Title page: *The great white shark is perhaps the ocean's most perfect predator.*

Editor: Frank Sloan

Cover and interior design by Nicola Stratford

Library of Congress Cataloging-in-Publication Data

Stone, Lynn M.
Oceans / Lynn M. Stone.
    p. cm. — (Biomes of North America)
Summary: Describes the landforms beneath the ocean's surface, marine plant and animal habitats, and how nature and people have changed the seas. Includes bibliographical references and index.
 ISBN 1-58952-686-4
 1. Marine ecology—Juvenile literature. 2. Ocean—Juvenile literature. [1. Ocean. 2. Marine ecology. 3. Ecology.] I. Title. II. Series: Stone, Lynn M. Biomes of North America.
 QH541.5.S3S693 2003
 577.7—dc21
                                            2003004600

Printed in the USA

CG/CG

Table of Contents

The Ocean	5
Ocean Communities	9
Ocean Animals	14
The Changing Ocean	19
Glossary	23
Index	24
Further Reading/Websites to Visit	24

Arctic Ocean
Greenland
(Den.)
Icelan
Alaska
(US)
Arctic Circle
Canada
United States
Pacific
Ocean
Atlantic
Ocean
Gulf of Mexico

The Ocean

    Two of every three parts of the Earth's surface are made up of oceans and seas. We have names for many oceans and seas. But they are all connected in one huge body of salt water.

    Because it is so huge, the ocean is very different from one place to another. Under the surface, for example, the ocean has grassy meadows, rocky **coral reefs**, canyons, mountains, sand, and mud.

The ocean is very shallow where sea grasses grow. But it can also be very deep. In the Pacific Ocean lies a trench 36,000 feet (10,973 meters) down. That is nearly 7 miles (11 kilometers) below the surface of the sea!

More than two-thirds of the Earth's surface is ocean, which shows many sides, from smooth to stormy, from warm to icy.

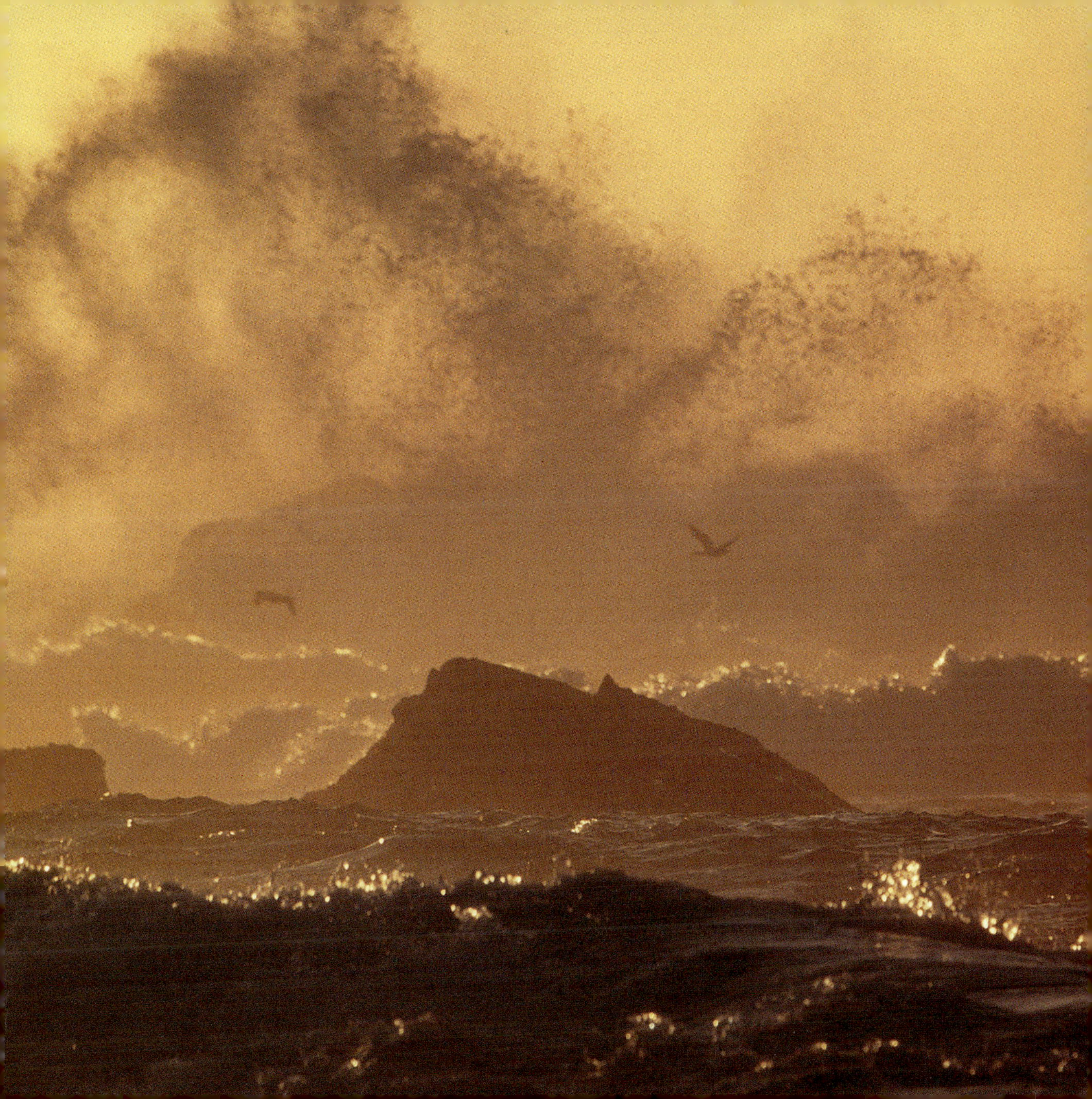

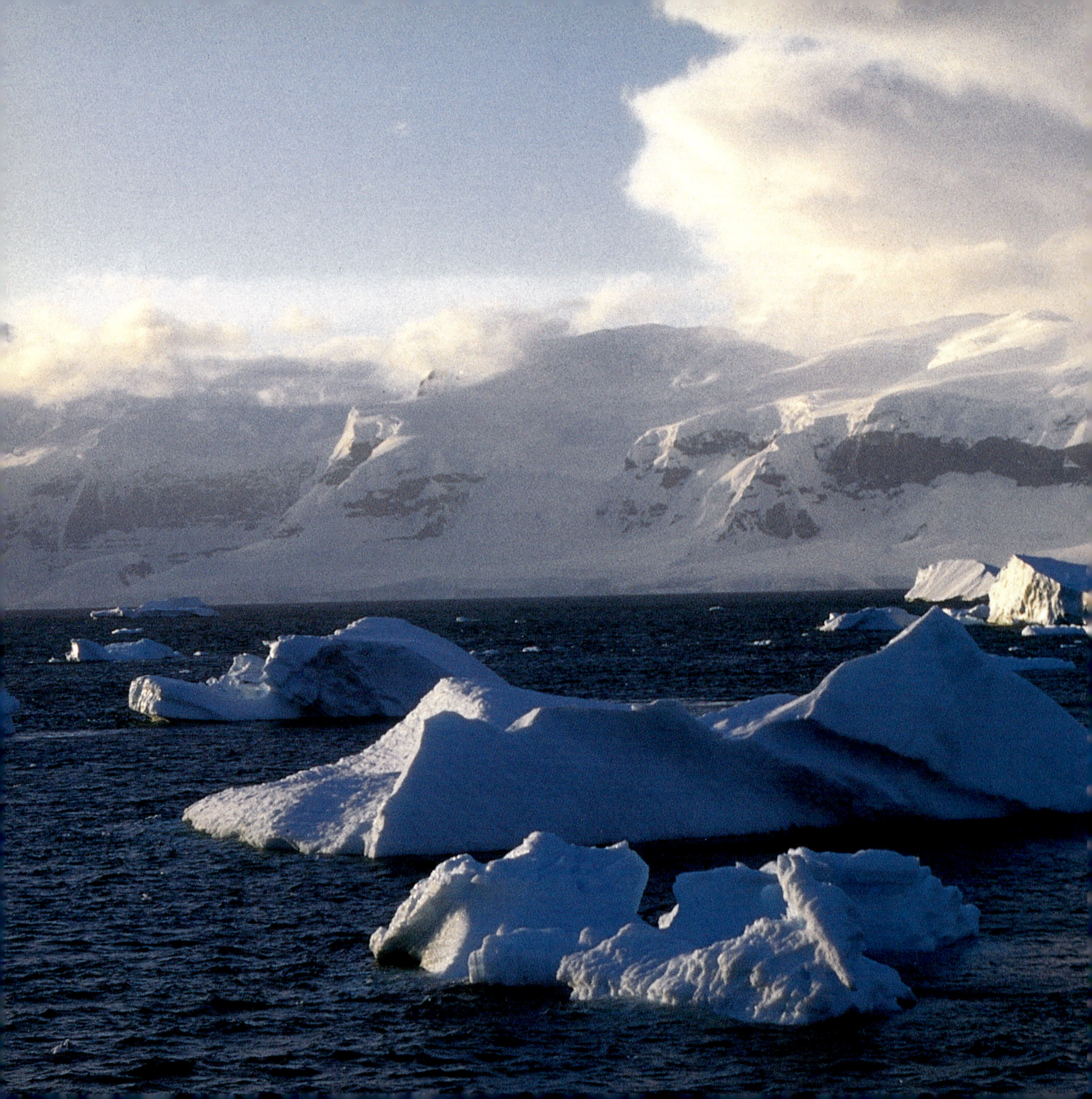

Ocean Communities

The ocean has many different communities of living things. These communities exist in many different ocean, or marine, habitats. Habitats are homes for plants and animals.

Each community is shaped by many forces. One force is **climate**. Climate is one thing that makes ocean water warm or cold.

Icebergs float in the oceans of the Arctic and Antarctic seas. But in the Caribbean Sea, the ocean water feels like bathwater.

Climate shapes many ocean communities, including those in the icy waters near Antarctica.

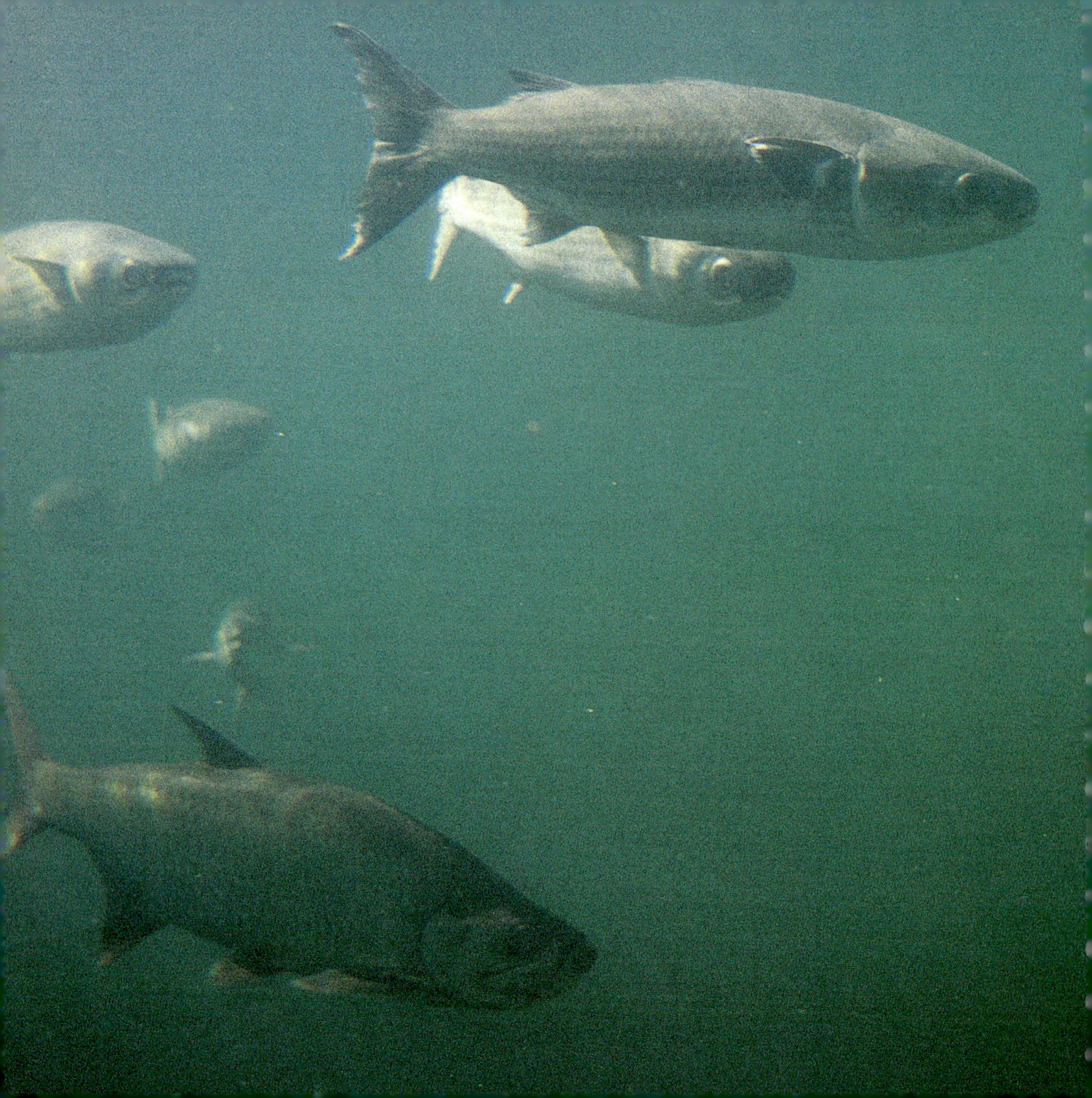

Certain animals and plants that live in warm water cannot live in cold water. Warm water communities have different kinds of fish than cold water communities. Most kinds of sharks, for example, like mild water temperatures. The Greenland shark, however, lives in icy Arctic seas.

Mullet (top) and tarpon live in warm, shallow seas and ocean bays.

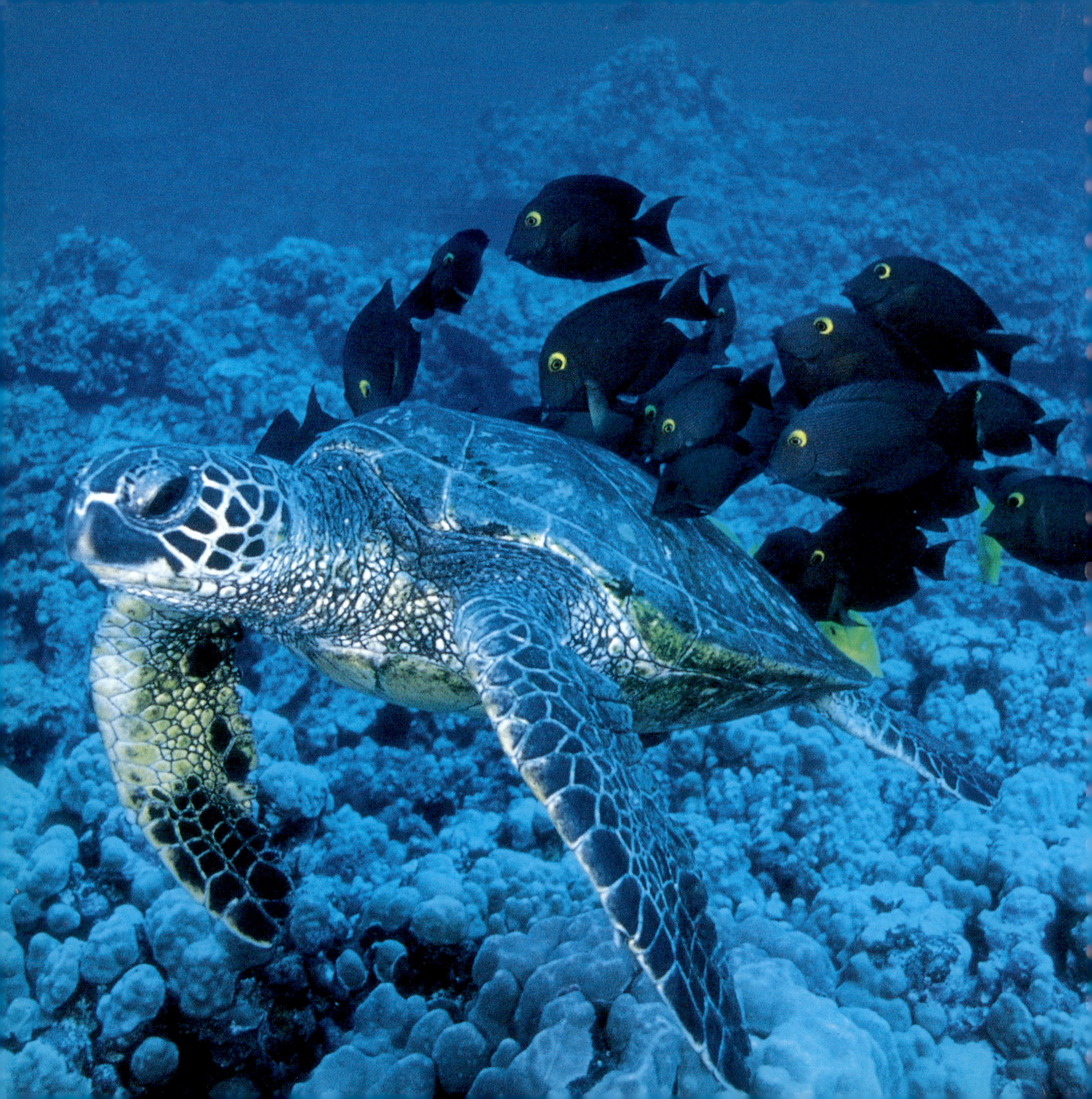

Water depth also shapes **marine** communities. Sunlight cannot reach deep water. Deep water creatures, like the lantern fish, live in darkness.

Since plants need sunlight to make food, they live in fairly shallow water. **Plankton** are tiny plants that are a basic food source of sea animals. Most marine animals, then, usually live in fairly shallow water.

Above a coral reef, cleaner fish follow a green sea turtle.

Ocean Animals

Ocean animals of one kind or another can be found throughout the oceans. Even the frigid Antarctic seas throb with sea life. Animals live on the surface and in the depths. About 9 of 10 ocean species live on the ocean bottom! Many of these bottom dwellers are bound to a single habitat, even a single rock! **Sea Stars**, and **shellfish**, like oysters, can travel only short distances.

Many marine animals survive through **camouflage**. Their colors match their surroundings.

The amazing octopus camouflages itself by changing color to blend in with the sea star behind it.

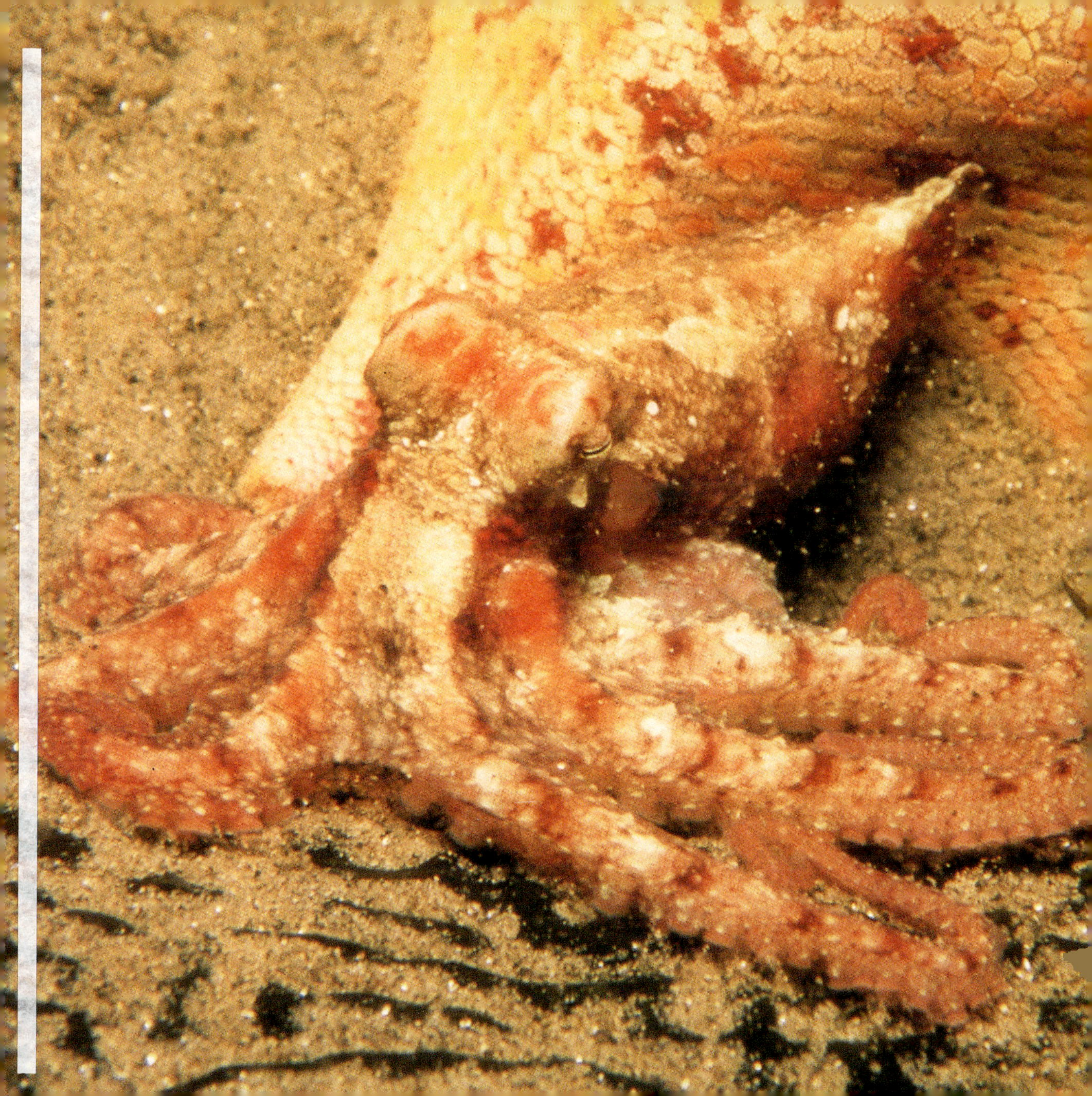

Other creatures, such as some kinds of fish, move from one habitat to another. In single dives, some kinds of whales and seals plunge from ocean light to ocean darkness. And certain whales and seals, along with certain fish, **migrate** long distances.

The giant sperm whale has been known to dive 10,500 feet (3,200 m) into the deep!

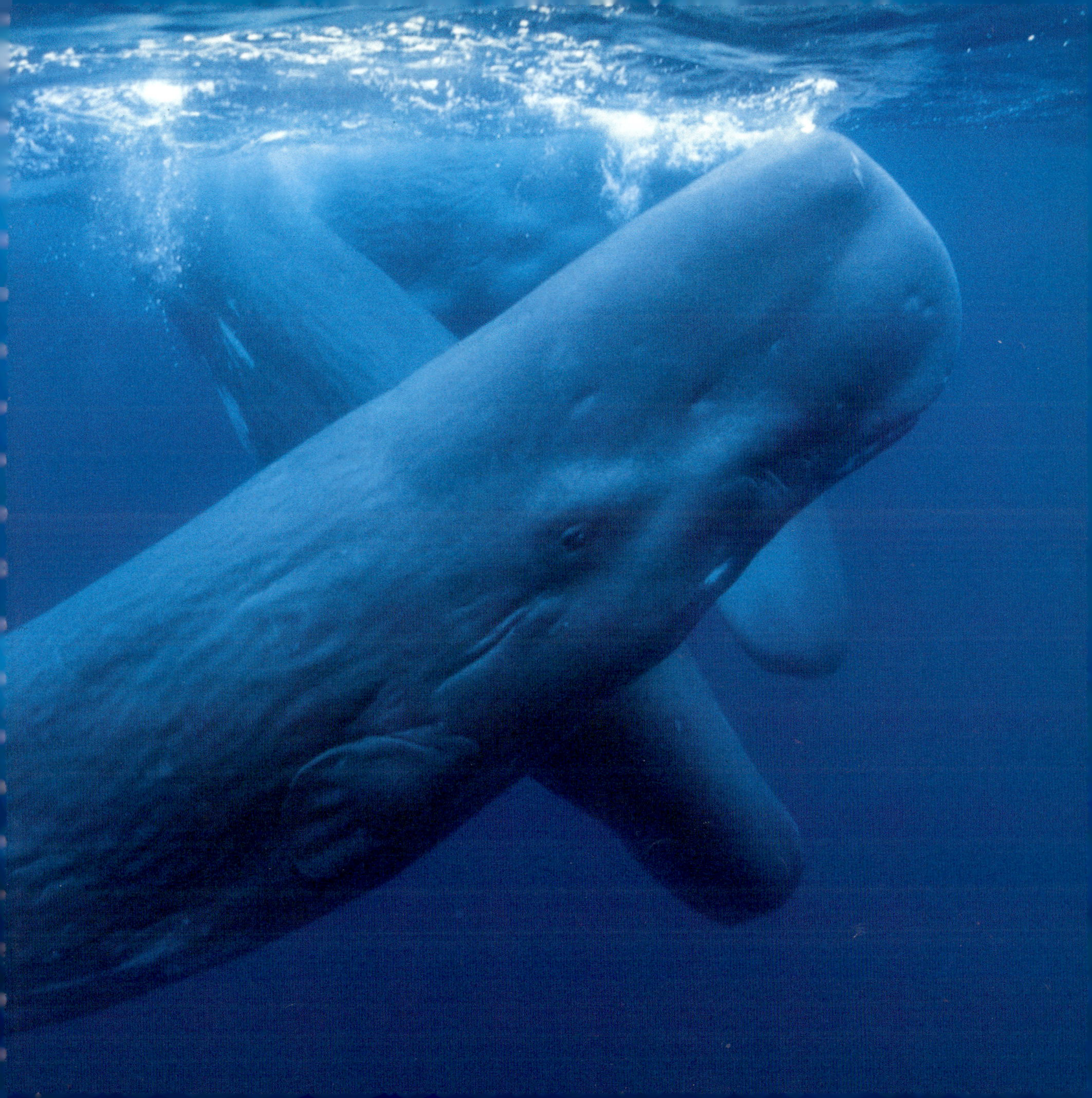

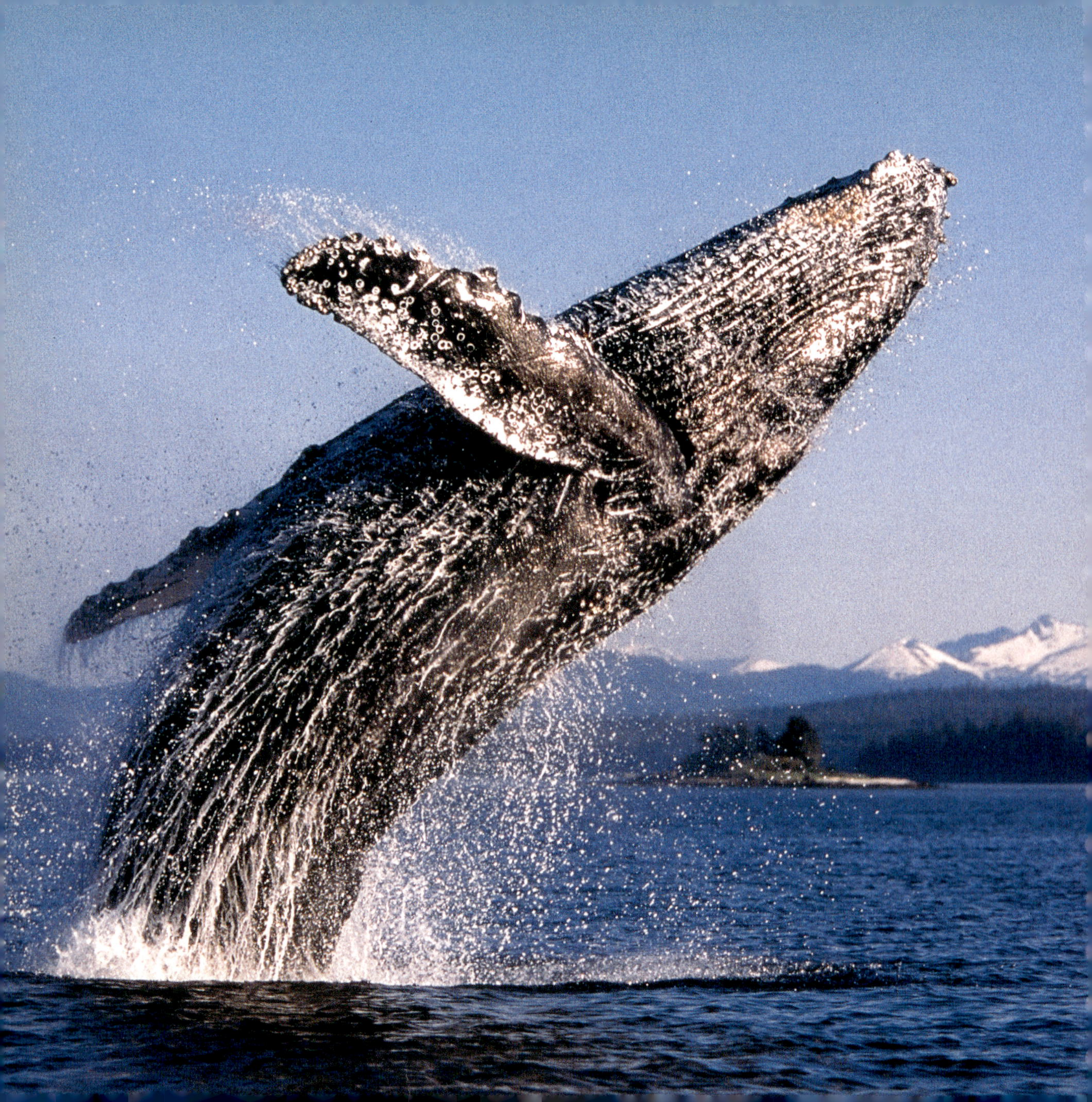

The Changing Ocean

Nature changes the ocean. The climates of different seasons change water temperatures. That forces some ocean animals to migrate. Others disappear for a time, hidden in sand or mud.

The climate change called **El Niño** causes a change in ocean currents. As currents change, some of the marine life also changes. Undersea volcanoes may heat up water temperatures and change the shape of the nearby ocean bottom.

Humpback whales migrate from cold seas into shallow, tropical seas during the breeding season.

In addition, people have changed the oceans by taking too much out—and putting too much in. Overfishing has caused fish like cod and swordfish to become scarce.

Meanwhile, people have added **pollutants** like chemicals, spilled oil, garbage, and old fishing nets to the ocean.

Fish eaters, like Steller's sea lions, now rare, appear to be hard hit from the lack of food. And killer whales, with fewer sea lions to eat, now eat more sea otters.

This Steller's sea lion has become tangled in an unwanted fishing net, which will probably kill the sea lion.

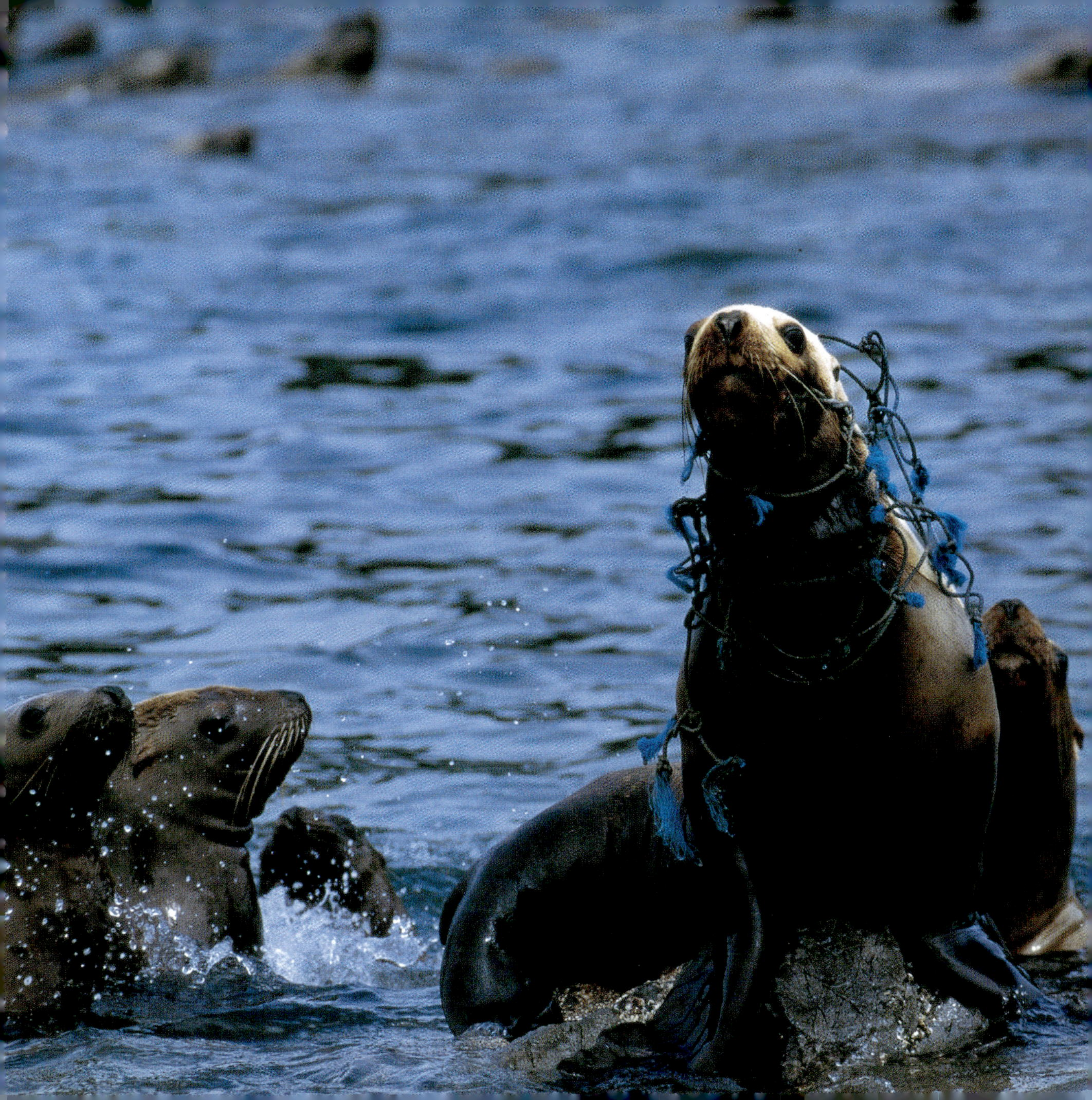

Ocean communities are a source of food for many animals that do not actually live in the sea, like bald eagles, seabirds, and humans.

Glossary

camouflage (KAM uh flazh) — the patterns of colors or shapes that allow an animal to blend into its surroundings

climate (KLY muht) — the weather of an area over a long period of time

coral reefs (KOR uhl REEFZ) — huge undersea structures naturally made by the limestone skeletons of certain coral animals

El Niño (ELL NEEN yo) — a change in ocean currents and temperature that causes changes in climate

marine (MUH reen) — of or relating to the sea

migrate (MY grayt) — to make a lengthy journey from one place to another at about the same time each year, usually spring and fall

plankton (PLANK tun) — mostly tiny, floating plants and animals of many kinds

pollutants (puh LOOT unts) — substances or things that poison or damage a natural habitat

sea stars (SEE STARZ) — kinds of sea creatures with "arms" leading forth from a central area; another name for starfish

shellfish (SHELL FISH) — not fish at all, but the common hard-shelled clams, oysters, crabs, lobsters, and mussels

INDEX

Antarctic Sea   9, 14
Arctic Sea   9, 11
camouflage   14
Caribbean Sea   9
climate   9, 19
coral reefs   5
El Niño   19
habitats   9, 14, 16
migrate   16, 19
Pacific Ocean   6
plankton   13
pollutants   20
salt water   5
sea grasses   6
sharks   11
Steller's sea lions   20

Further Reading

Baker, Lucy. *Life in the Oceans*. Creative Publications, Intl., 2000
Gray, Susan. *Oceans*. Compass Point Books, 2001
MacQuitty, Miranda. *Ocean*. DK Publishing, 2000
Steele, Christy. *Oceans*. Steadwell Books, 2001

Websites To Visit

mbgnet.mobot.org/salt/index.htm
dir.yahoo.com/Society_and_Culture/Environment_and_Nature/Water_Resources/
    Oceans_and_Seas

About The Author

Lynn Stone is a talented natural history photographer and writer. Lynn, a former teacher, travels worldwide to photograph wildlife in their natural habitat. He has more than 500 children's books to his credit.

THE CHICAGO PUBLIC LIBRARY